Global Symposium on Leadership and Project Management 2024

AA004030

Online
22-23 August 2024

ISBN: 979-8-3313-0142-2
eISBN: 979-8-3313-0143-9

Printed from e-media with permission by:

Curran Associates, Inc.
57 Morehouse Lane
Red Hook, NY 12571

Some format issues inherent in the e-media version may also appear in this print version.

Phone: (857) 301-0010

pmsymposium@cps.northeastern.edu

Additional copies of this publication are available from:

Curran Associates, Inc.
57 Morehouse Lane
Red Hook, NY 12571 USA
Phone: 845-758-0400
Fax: 845-758-2633
Email: curran@proceedings.com
Web: www.proceedings.com

TABLE OF CONTENTS

An Empirical Analysis of Agile Manifesto

Pedro Serrador

College of Professional Studies, Northeastern University, Toronto, ON, Canada
p.serrador@northeastern.edu

Abstract

Established in 2001, the Agile Manifesto has grown to impact many other fields in addition to software development practices. This research-based critique aims to examine the strengths and weaknesses of this influential document. This study utilized two large datasets to examine the strengths and weaknesses of the Agile Manifesto. Through an analysis of values and principles, methodologies, and empirical data, the study analyzes the validity of each of the values and principals from an empirical viewpoint. The findings suggest the need for a review and revision of the manifesto to align it with the evolving industry landscape.

Keywords: agile manifesto, agile methodologies, project success, project complexity, agility

1. Introduction

Published in 2001, the Agile Manifesto was released by a group of 17 software process methodologists, who attended a summit to promote a better way of developing software and then formed the Agile Alliance (Fitsilis et al., 2014; Vierlboeck et al., 2019). Since the creation of the Agile manifesto in 2001, it has revolutionized the way software development projects are managed and it has been growing vertically within organizations and horizontally across organizations in different business sectors (Rahy & Bass, 2022). The Agile Manifesto was created in response to the Waterfall framework to create a new foundation for frameworks that are better able to cope with increasing uncertainties (Sandstø & Reme-Ness, 2021).

The Agile Manifesto has not only gained traction but has become the predominant methodologies in use (Kinnunen & Luoma, 2018). Agile principles and their implementation in practice have paved the way for radical new and innovative ways of software and product development (Hohl et al., 2018). Companies often take recourse to Agile software development methods in a distributed environment in search of reduced cost, higher efficiency, increased flexibility, and good customization (Bose, 2008). Since the statement of the Agile Manifesto, Agile methods are being studied by the scientific community and by the software industry, which is increasingly promoting their use in software projects (Tolfo et al., 2018). The concept of Agile methods appeared at the beginning of the millennium with the launching of the Agile manifesto in 2001 (Deneckere et al., 2016). In this environment, independently of the business area, we see that more and more methods for specification and software development are being gradually replaced or upgraded by the principles and values announced in the Agile Manifesto (Luna et al., 2010).

Empirical studies, such as the research by Serrador and Pinto in 2015, consistently reveal higher success rates in Agile projects (Serrador & Pinto, 2015). Beyond the confines of software development, Agile methodologies find extensive adoption across diverse industries, underscoring their versatile applicability (Looks et al., 2021; Salinas et al., 2018). Notably, within the software development landscape, Agile processes have not merely gained traction

 https://doi.org/10.52202/075831-0001

but have become the predominant methodologies in use (Hohl et al., 2018; Kuhrmann et al., 2022).

Despite the successes of the Agile Manifesto, it is important to realize that there are areas where it is not perfect or could be improved. Studies have revealed that Agile processes are frequently used in a modified format known as "Hybrid Agile." In fact, Hybrid Agile methods constitute the largest percentage of projects utilizing Agile methodologies (Gemino et al., 2021). This observation alone makes a compelling argument for revisiting the Agile Manifesto and considering possible updates or adjustments to better adapt it to the realities and complexities of modern project environments. By recognizing the prevalence of Hybrid Agile and the need for continuous improvement, we can ensure that Agile methodologies effectively address the evolving needs and challenges of modern projects.

Agile methodology is based on four key values: individuals and interactions over processes and tools, working software over comprehensive documentation, customer collaboration over contract negotiation, and responding to change over following a plan (Looks et al., 2021). These values emphasize the importance of flexibility, collaboration, and adaptability in software development processes (Schmitt & Hörner, 2021).

These Agile values emphasize the importance of adaptability, collaboration, and customer-centricity in software development (Madi et al., 2011). By prioritizing individuals and interactions, Agile methodologies aim to create a culture of collaboration and effective communication within project teams (Cockburn & Highsmith, 2001). The emphasis on working software over comprehensive documentation promotes the delivery of tangible results to customers and stakeholders (Behutiye et al., 2022). The value of customer collaboration over contract negotiation highlights the need to actively involve customers throughout the development process to ensure their needs and expectations are met (Dingsøyr et al., 2012). Lastly, the focus on responding to change over following a plan emphasizes the importance of adaptability and flexibility in an ever-changing project landscape (Kantola et al., 2022).

Hohl et al. (2018) interviewed a number of the original authors of the Manifesto and found that though some authors were in favor of a revision, they were not in the majority and most did not believe an update was warranted. Beck, one of the original authors of the manifesto, proposed an update of the Agile Manifesto in 2010. His proposed values were (Denning, 2015):

- Team vision and discipline/over individuals and interactions/over processes and tools.
- Validated learning/over working software /over comprehensive documentation.
- Customer discovery/over customer collaboration/over contract negotiation.
- Initiating change/over responding to change/over following a plan.

However, this revision has not gained traction. Perhaps what is needed is a less radical and more evidence-based review.

In this study, the values and principles outlined in the Agile Manifesto will be treated as hypotheses that will be subjected to examination. The aim is to gather quantitative data that can provide support for these hypotheses. If quantitative data is not readily available, a concise discussion will be included to further explore the relationships between the values and

https://doi.org/10.52202/075831-0001

principles of Agile methodologies. By adopting this approach, we hope to enhance the rigor and validity of our analysis.

According to the Agile Manifesto (Beck et al., 2001), There are four key values:

V1. Individuals and interactions over processes and tools
V2. Working software over comprehensive documentation
V3. Customer collaboration over contract negotiation
V4. Responding to change over following a plan

And 12 principals:

P1. Our highest priority is to satisfy the customer through early and continuous delivery of valuable software.
P2. Welcome changing requirements, even late in development. Agile processes harness change for the customer's competitive advantage.
P3. Deliver working software frequently, from a couple of weeks to a couple of months, with a preference to the shorter timescale.
P4. Business people and developers must work together daily throughout the project.
P5. Build projects around motivated individuals. Give them the environment and support they need and trust them to get the job done.
P6. The most efficient and effective method of conveying information to and within a development team is face-to-face conversation.
P7. Working software is the primary measure of progress.
P8. Agile processes promote sustainable development. The sponsors, developers, and users should be able to maintain a constant pace indefinitely.
P9. Continuous attention to technical excellence and good design enhances agility.
P10. Simplicity--the art of maximizing the amount of work not done--is essential.
P11. The best architectures, requirements, and designs emerge from self-organizing teams.
P12. At regular intervals, the team reflects on how to become more effective, then tunes and adjusts its behavior accordingly.

We will treat these values and principles as hypotheses that we will examine. We will look for quantitative data that could be used to support these hypotheses and, where not available, we will have a short discussion on each one.

2. Methodology

The section outlines the approach taken to analyze the relationship between Agile methodologies and project success. This research aims to examine how Agile values and principles have been successful in achieving promised solutions, including project efficiency, stakeholder success, and project success (See Figure 1). By evaluating the effectiveness of Agile methodologies, this study seeks to understand the impact of Agile values and principles on various project outcomes. Specifically, it explores the extent to which Agile methodologies contribute to improved project efficiency, ensuring that projects are executed in a timely and cost-effective manner, stakeholder success, which involves satisfying the needs and expectations of stakeholders throughout the project lifecycle and finally, the overall project success reported by respondents. Through this investigation, the study aims to provide valuable

 https://doi.org/10.52202/075831-0001

insights into the effectiveness of Agile methodologies in achieving desired project outcomes. Due to the challenges of large-scale data collection in the current landscape, the analysis relies on re-analyzing data from two previous studies on project success. The objective is to explore if these studies provide insights that can address the research questions at hand. Should the findings prove interesting, it is anticipated that further research may be conducted to delve deeper into the topic.

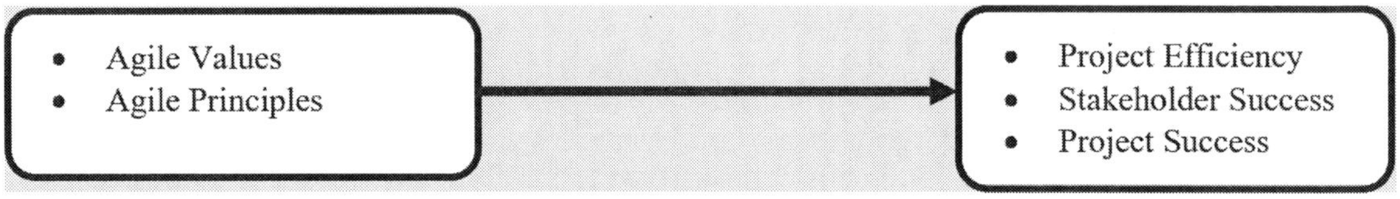

Figure 1 Principal research constructs

Data was collected on various parameters, including success metrics, the utilization of Agile methodologies, and project characteristics influencing success. A quantitative methodology was employed to analyze this dataset. To assess perceived project success, survey questions were formulated based on a composite of success dimensions outlined by Müller and Turner (2007), Shenhar et al. (2001), and Crawford et al. (2004). These success measures are well-delineated in the existing literature and have been incorporated into the survey questionnaire. Overall, we follow the success measures as defined in (Serrador & Turner, 2015).

Two datasets were examined in this study. Dataset 1 consisted of survey data from 477 software and non-software projects across industries and countries. After data cleaning, 449 projects remained for analysis. Dataset 2 consisted of a larger dataset of 1,539 projects, of which 1,002 had sufficient information for analysis from an Agile perspective. Dataset 1 was specifically designed to investigate Agile success and will be used for primary analysis, while Dataset 2 will be utilized for additional analysis where appropriate. Both datasets have been analyzed and used in previously published research on Agile and Hybrid Agile projects. Both datasets were analyzed for normality and success measures were analyzed using factor analysis and Cronbach Alpha for internal validity.

This study adopts a post-positivist perspective, positing a correlation between the extent of project planning and the achievement of project success. Post-positivism resides between positivism, which assumes the attainability of wholly objective research outcomes, and phenomenology, which emphasizes subjective experiences. Recognizing the subjective nature of perception and observation, it is acknowledged that results cannot be entirely objective. The concept of project success, for instance, may resist complete quantification and is subject to the subjective assessments of participants and sponsors. Hence, the epistemological approach employed in this study is post-positivism, acknowledging the value of insights from positivism while recognizing its limitations in the realm of business research. By adopting this methodology, this study seeks to shed light on the relationship between Agile methodologies and project success, leveraging existing very large dataset. No study approaching this scale has been completed previously.

The values and principles espoused by Agile methodologies present challenges in their quantitative analysis due to their general nature and impact on multiple aspects of a project. To address this issue, a focused approach was taken in this study, examining subsets of documentation rather than attempting to analyze the broad category as a whole. Although specific references supporting this approach could not be identified, it is important to note that researchers often adopt targeted analyses to explore complex phenomena comprehensively. For this study, a straightforward method of analysis was employed to provide an initial review of the relationships under study. An analysis of variance (ANOVA) was conducted to compare the means between less successful and more successful projects, providing valuable insights into the relationship between variables. This approach offered a preliminary understanding of the likely existence of relationships. In instances where appropriate, linear regression analysis was performed to further explore the dataset. It is worth noting that future research endeavors could adopt more advanced statistical methods to enhance the analytical rigor and sophistication of their analyses, ensuring a deeper understanding of the relationships under investigation.

By employing a combination of ANOVA and linear regression analyses, this study has provided valuable insights into the relationship between Agile values, principles, and project success. However, future research can build upon these preliminary findings by applying more sophisticated statistical techniques tailored to the specific research questions at hand. This would contribute to a deeper understanding of Agile methodologies and their impact on project management and software development.

3. Hypotheses

For the purpose of this research, we will treat each of the 4 values and each of the 12 principles as hypotheses. This provides a total of 16 hypotheses to be tested.

In the subsequent analysis, we will systematically examine the four core values (V1-4) and twelve principles (P1-12) outlined in the Agile Manifesto, aiming to examine the relevant quantitative data that can either support or challenge these hypotheses. Our objective is to critically assess the empirical evidence and determine the extent to which these values and principles contribute to project success in the context of Agile methodologies.

3.1 Agile Values

3.1.1 V1. Individuals and interactions over processes and tools

Is Agile not a process? This value statement is extremely difficult to verify from a quantitative perspective. At first glance, this statement seems reasonable, as many people believe that overuse of processes and tools can hinder project progress. Anecdotes from practitioners often report cases of project failures due to an overemphasis on the process. Additionally, research indicates that projects that dedicate excessive time to process and planning exhibit lower success rates (Serrador & Turner, 2015).

However, it is worth noting that Scrum, the most popular Agile methodology, is a process itself (Kadenic et al., 2023). This raises the question of how one can accurately measure the impact of prioritizing process over team emphasis. While too much focus on process and tools can be detrimental, disregarding them entirely may also lead to unfavorable outcomes (Salinas et al., 2018). Insufficient planning and process implementation have been linked to project failures (Serrador & Turner, 2015). In light of these complexities, it is difficult to definitively verify or refute this statement. Consequently, a lack of compelling evidence exists to support either perspective. Further research and empirical studies are needed to explore this relationship in more depth.

V1: We cannot confirm that V1 is associated with more or less project success.

 https://doi.org/10.52202/075831-0001

3.1.2 V2 - Working software over comprehensive documentation

The Agile Manifesto emphasizes the value of working software over comprehensive documentation. While it is evident that the primary focus should be on delivering functional software to customers, the importance of comprehensive documentation is sometimes overlooked. The ANOVA results presented in Table 1 and

Table 2 provide insights into the relationship between the quality of non-functional requirements documentation and project success. unless noted, data analysis provided in tables are from dataset 1. In the more Agile projects, the means of success measures generally increase as the quality of non-functional requirements documentation improves. For instance, the mean stakeholder success scores range from 0.596 for the "Very Low/Not Used" category to 0.887 for the "Very High" category. The p-values (P(F)) for all success measures are significant ($p < 0.001$), indicating a statistically significant relationship between the quality of non-functional requirements documentation and project success.

Similarly, in the analysis including all groups, higher means are observed for success measures as the quality of non-functional requirements documentation improves. For example, the mean stakeholder success scores range from 0.591 for the "Very Low/Not Used" category to 0.846 for the "Very High" category. The p-values (P(F)) are significant ($p < 0.001$) for all success measures, further indicating the statistically significant relationship between the quality of non-functional requirements documentation and project success.

These findings highlight the importance of comprehensive and high-quality non-functional requirements documentation in achieving higher levels of project success. As the quality of documentation improves, the project's success measures show corresponding improvements. These results underscore the need for organizations to prioritize and invest in the development of strong documentation practices to enhance project outcomes.

Table 1 ANOVA Quality of Non-functional requirements document and Success (More Agile)

Quality of non-functional requirements documentation	Means and ANOVA				
	Success stakeholder	Success overall	Success budget/time	Success scope/quality	Valid N
Very Low/not used	0.596	0.594	0.520	0.663	18
Low	0.677	0.650	0.576	0.670	46
Medium	0.731	0.719	0.676	0.738	95
High	0.767	0.766	0.740	0.787	47
Very High	0.887	0.845	0.757	0.851	10
Total	0.723	0.710	0.659	0.733	216
P(F)	**0.000**	**0.000**	**0.000**	**0.002**	

Table 2 ANOVA Quality of Non-functional requirements document and Success (All Groups)

Quality of non-functional requirements documentation	Means and ANOVA				
	Success stakeholder	Success overall	Success budget/time	Success scope/quality	Valid N
Very Low/not used	0.591	0.592	0.535	0.651	47

Low	0.653	0.640	0.595	0.658	113
Medium	0.707	0.707	0.672	0.741	175
High	0.735	0.733	0.692	0.768	95
Very High	0.846	0.815	0.723	0.844	16
Total	0.692	0.687	0.644	0.720	446
P(F)	**0.000**	**0.000**	**0.000**	**0.000**	

Another type of common documentation in projects is the WBS, the work breakdown structure. The detail level of this documentation was also measured against success.

Table 3 ANOVA Success and Detail Level of WBS (Less Agile)

Detail level of WBS	Means and ANOVA				
	Success stakeholder	Success overall	Success budget/time	Success scope/quality	Valid N
Very Low/not used	0.569	0.580	0.546	0.637	17
Low	0.591	0.614	0.624	0.652	50
Medium	0.679	0.682	0.647	0.721	101
High	0.715	0.705	0.634	0.756	47
Very High	0.720	0.700	0.648	0.713	15
Total	0.662	0.665	0.632	0.707	230
P(F)	**0.004**	**0.018**	0.576	**0.030**	

Table 4 ANOVA Success and Detail Level of WBS (More Agile)

Detail level of WBS	Means and ANOVA				
	Success stakeholder	Success overall	Success budget/time	Success scope/quality	Valid N
Very Low/not used	0.657	0.621	0.537	0.633	27
Low	0.665	0.657	0.609	0.689	40
Medium	0.732	0.728	0.682	0.765	78
High	0.795	0.767	0.716	0.764	55
Very High	0.696	0.707	0.688	0.747	16
Total	0.724	0.710	0.659	0.733	216
P(F)	**0.002**	**0.001**	**0.003**	**0.005**	

Table 5 ANOVA Success and Detail Level of WBS (All Groups)

Detail level of WBS	Means and ANOVA				
	Success stakeholder	Success overall	Success budget/time	Success scope/quality	Valid N
Very Low/not used	0.623	0.605	0.541	0.634	44
Low	0.624	0.633	0.617	0.669	90
Medium	0.702	0.702	0.662	0.740	179

https://doi.org/10.52202/075831-0001

High	0.758	0.739	0.678	0.760	102
Very High	0.708	0.704	0.668	0.731	31
Total	0.692	0.687	0.645	0.719	446
P(F)	**0.000**	**0.000**	**0.006**	**0.000**	

Our analysis (Table 3-Table 5) revealed a clear association between the detail level of the WBS and project success. This association held true for all aspects of success and, unexpectedly, appeared to have a stronger impact on more Agile projects compared to less Agile projects. It is worth noting that the detailed WBSs were helpful in driving success across all measures.

Interestingly, while the detail level of the WBS was associated with success in various aspects, it was not significantly correlated with time and budget success for less Agile projects. This finding suggests that while a detailed WBS may not directly contribute to time and budget success in less Agile projects, it remains beneficial in other areas of project success.

We now do a further analysis using regression to confirm our findings. We can see from Table 6 -
Table 8, that there is a consistently strong relationship between the quality of documentation and project success across success measures. This confirms the ANOVA analysis already performed.

Table 6 Regression of Functional Requirements quality and success

Success and quality of requirements	R^2	P	Valid N
Success overall	.110	.000	448
Success scope/quality	.094	,000	448
Success budget/time	.057	.000	448

Table 7 Regression of non-Functional Requirements quality and success

Success and quality of requirements	R^2	P	Valid N
Success overall	.074	.000	448
Success scope/quality	.072	.000	448
Success budget/time	.050	.000	448

Table 8 Regression of WBS quality and success

Success and quality of requirements	R^2	P	Valid N
Success overall	.050	.000	448
Success scope/quality	.037	.000	448
Success budget/time	.022	.002	448

Considering the importance of documentation to Agile projects, one could argue that the statement in V2 may be misleading. It is clear from our research that proper documentation, including comprehensive non-functional requirements and detailed WBSs, is crucial in maximizing project success across all projects. Therefore, it may be more appropriate to reword V2 as "Working software rather than overly detailed documentation," highlighting the

https://doi.org/10.52202/075831-0001

importance of avoiding excessive and unnecessary documentation rather than dismissing the need for documentation altogether.

V2: We cannot confirm that V2 is associated with enhanced project success. In fact, the opposite appears to be true.

3.1.3 V3 - Customer collaboration over contract negotiation

To examine the statement of value regarding customer collaboration over contract negotiation (V3), we analyzed the level of stakeholder engagement in project processes and its correlation with project success.

The findings of the regression analysis (Table 9) indicate that stakeholder engagement is significantly associated with project success, both for more Agile and less Agile projects. The degree of stakeholder engagement explains approximately 22% to 25% of the variance in success. Additionally, the ANOVA results (Table 10-Table 12) show that higher levels of stakeholder engagement are correlated with higher levels of success across all measures. This relationship holds true for all types of projects, including those using Scrum methodology (Table 13). Scrum or non-scrum. In-fact, R^2 is higher for non-scrum projects, indicating it is more important in the non-Scrum projects. This may be because, scrum as part of the process, ensures stakeholder are closely involved.

Table 9 Regression Degree of Stakeholder Engagement and Stakeholder Success

Stakeholder success and stakeholder engagement	R^2	P	Valid N
More Agile	0.219	0.000	217
Less Agile	0.252	0.000	231
All Groups	0.251	0.000	448

Table 10 ANOVA Degree of Stakeholder Engagement and Success (Less Agile)

Degree of stakeholder engagement	Means and ANOVA				
	Success stakeholder	Success overall	Success budget/time	Success scope/quality	Valid N
Very Low/not used	0.400	0.443	0.507	0.463	10
Low	0.553	0.564	0.536	0.616	34
Medium	0.625	0.635	0.612	0.678	79
High	0.709	0.709	0.657	0.762	82
Very High	0.864	0.835	0.786	0.826	26
Total	0.661	0.665	0.632	0.706	231
P(F)	**0.000**	**0.000**	**0.000**	**0.000**	

Table 11 ANOVA Degree of Stakeholder Engagement and Success (More Agile)

Degree of	Means and ANOVA

https://doi.org/10.52202/075831-0001

stakeholder engagement	Success stakeholder	Success overall	Success budget/time	Success scope/quality	Valid N
Very Low/not used	0.500	0.371	0.179	0.307	2
Low	0.512	0.532	0.474	0.629	22
Medium	0.678	0.668	0.627	0.690	60
High	0.763	0.747	0.704	0.759	89
Very High	0.820	0.791	0.716	0.808	44
Total	0.723	0.709	0.657	0.733	217
P(F)	**0.000**	**0.000**	**0.000**	**0.000**	

Table 12 ANOVA Degree of Stakeholder Engagement and Success (All Groups)

Degree of stakeholder engagement	Means and ANOVA				
	Success stakeholder	Success Overall	Success budget/time	Success scope/quality	Valid N
Very Low	0.417	0.431	0.452	0.437	12
Low	0.537	0.552	0.511	0.621	56
Medium	0.648	0.649	0.619	0.683	139
High	0.737	0.729	0.681	0.760	171
Very High	0.836	0.807	0.742	0.815	70
Total	0.691	0.686	0.644	0.719	448
p(F)	**0.000**	**0.000**	**0.000**	**0.000**	

Table 13 Regression Stakeholder Engagement and Success for Projects Using and not Using SCRUM

	Success overall	Stakeholder Success	Scope/quality Success	Success budget/time
SCRUM				
R^2	0.176	0.185	0.102	0.112
P	0.000	0.000	0.000	0.000
Valid N	210			
NON-SCRUM				
R^2	0.290	0.298	0.226	0.100
P	0.000	0.000	0.000	0.000
Valid N	238			

In summary, these results provide strong evidence that stakeholder engagement is factor in achieving project success. Engaging stakeholders throughout the project lifecycle can contribute to improved outcomes and better alignment with project objectives.

V3: We can therefore confirm that V3 is associated with enhanced project success.

https://doi.org/10.52202/075831-0001

3.1.4 V4 - Responding to change over following a plan

The analysis of team attitude toward accepting changes and project success reveals significant findings. The ANOVA results demonstrate that higher levels of team attitude toward accepting changes are associated with higher levels of success across various measures (Table 14-Table 16). For example, in less Agile projects, the mean success scores increase from 0.433 for "very low" team attitude to accepting changes to 0.778 for "very high". This is to be expected as being too flexible in that situation can lead to increased timelines and budget. For more Agile projects, it is strongly associated with success including budget/time success. The p-values are all 0.000, indicating a strong statistical significance. These results suggest that embracing change and being adaptable positively impact project success.

However, the analysis also indicates that budget flexibility does not significantly contribute to project success (Table 17- Table 19). The ANOVA results show no significant association between budget flexibility and success, except for a negative correlation with success in budget/time for more Agile projects.

Table 14 ANOVA Team Attitude toward Accepting Changes and Success (Less Agile)

Team attitude toward accepting changes	Means and ANOVA				
	Success stakeholder	Success Overall	Success budget/time	Success scope/quality	Valid N
Very Low	0.433	0.492	0.550	0.550	10
Low	0.558	0.583	0.589	0.626	48
Medium	0.663	0.664	0.620	0.713	96
High	0.749	0.736	0.681	0.764	61
Very High	0.778	0.753	0.667	0.789	15
Total	0.661	0.665	0.630	0.706	230
p(F)	**0.000**	**0.000**	**0.180**	**0.000**	

Table 15 ANOVA Team Attitude toward Accepting Changes and Success (More Agile)

Team attitude toward accepting changes	Means and ANOVA				
	Success stakeholder	Success Overall	Success budget/time	Success scope/quality	Valid N
Very Low					
Low	0.538	0.544	0.490	0.610	15
Medium	0.680	0.666	0.614	0.689	80
High	0.778	0.757	0.703	0.769	89
Very High	0.764	0.759	0.712	0.797	33
Total	0.723	0.709	0.657	0.733	217
p(F)	**0.000**	**0.000**	**0.000**	**0.000**	

Table 16 ANOVA Team Attitude toward Accepting Changes and Success (All Groups)

	Means and ANOVA

 https://doi.org/10.52202/075831-0001

Team attitude toward accepting changes	Success stakeholder	Success Overall	Success budget/time	Success scope/quality	Valid N
Very Low	0.433	0.492	0.550	0.550	10
Low	0.553	0.574	0.566	0.622	63
Medium	0.670	0.665	0.617	0.702	176
High	0.766	0.748	0.694	0.767	150
Very High	0.768	0.757	0.698	0.794	48
Total	0.691	0.686	0.643	0.719	447
p(F)	**0.000**	**0.000**	**0.000**	**0.000**	

Table 17 ANOVA Budget Flexibility and Success (Less Agile)

Question on budget flexibility	Means and ANOVA				
	Success stakeholder	Success Overall	Success budget/time	Success scope/quality	Valid N
Very Low	0.686	0.691	0.664	0.729	31
Low	0.651	0.661	0.629	0.711	101
Medium	0.664	0.668	0.651	0.692	74
High	0.653	0.620	0.489	0.684	19
Very High	0.676	0.687	0.643	0.751	7
Total	0.661	0.664	0.630	0.706	232
p(F)	0.940	0.722	0.079	0.807	

Table 18 ANOVA Budget Flexibility and Success (More Agile)

Question on budget flexibility	Means and ANOVA				
	Success stakeholder	Success Overall	Success budget/time	Success scope/quality	Valid N
Very Low	0.698	0.710	0.735	0.710	21
Low	0.730	0.725	0.687	0.752	73
Medium	0.739	0.719	0.656	0.742	87
High	0.678	0.646	0.556	0.671	23
Very High	0.697	0.663	0.549	0.709	13
Total	0.723	0.709	0.657	0.733	217
p(F)	0.617	0.301	0.017	0.366	

*Negative correlation

Table 19 ANOVA Budget Flexibility and Success (All Groups)

Question on budget flexibility	Means and ANOVA				
	Success stakeholder	Success Overall	Success budget/time	Success scope/quality	Valid N
Very Low	0.691	0.699	0.692	0.721	52
Low	0.684	0.688	0.654	0.728	174
Medium	0.704	0.695	0.654	0.719	161
High	0.667	0.634	0.526	0.677	42
Very High	0.690	0.671	0.582	0.724	20

https://doi.org/10.52202/075831-0001

Total	0.691	0.686	0.643	0.719	449
p(F)	0.810	0.338	**0.003**	0.617	

*Negative correlation

The findings confirm that team attitude toward accepting changes, as reflected in higher levels of acceptance, is associated with enhanced project success. On the other hand, budget flexibility does not demonstrate a positive impact on project outcomes. Therefore, it is crucial for project teams to prioritize adaptability and change management while ensuring responsible budget practices.

V4: We can therefore confirm that V4 is associated with enhanced project success.

3.5 Principles behind the Agile Manifesto

We now take the same approach to review the 12 principles of the Agile Manifesto.

3.5.1 P1: Our highest priority is to satisfy the customer through early and continuous delivery of valuable software.

We conducted an analysis to assess the relationship between Agile use and stakeholder success factors, focusing on Principle 1 of the Agile Manifesto. This principle is similar to Value #1 so we can examine Table 1Table 5 for guidance here. The analysis conducted on both dataset 1 and 2 confirms that Agile use is associated with enhanced stakeholder satisfaction, supporting Principle 1 of the Agile Manifesto. If we take another approach, we find a clear relationship between Agile methodologies and success factors such as stakeholder satisfaction, overall success, budget/time success, and scope/quality success. There is a clear relationship between Agile use and success. P is significant in all cases. R2 is much higher for stakeholder success (.06) and lowest for budget, time (.01), see Table 20 and Table 21.

Table 20 Regression Agility Metric and Success

Agility Factor and Success	R^2	p
Success Stakeholder	0.061	0.000
Success Overall	0.046	0.000
Success Budget/time	0.013	0.015
Success Scope/quality	0.024	0.001
Valid N	449	

Table 21 Comparison for R^2 from MHRA analysis between Agile measures and success factors for dataset 2

N=1002	Stakeholder Success Factor	Efficiency factor
R^2	.152	.096
	p=.089	*p=.083*

The results indicate that Agile processes contribute to improved project outcomes and highlight the importance of prioritizing customer satisfaction through early and continuous delivery of

https://doi.org/10.52202/075831-0001

valuable software. These findings reinforce the value of adopting Agile methodologies in software development projects.
P1: We can therefore confirm that Agile use is associated with enhanced stakeholder satisfaction.

3.5.2 P2. Welcome changing requirements, even late in development. Agile processes harness change for the customer's competitive advantage.

As discussed in our analysis of V4, above, we asked the question:" How well did the team and the project respond to external changes?" The results of Table 14-Table 16 are applicable here also. Flexibility to changing requirements is associated with more successful projects. Table 14-Table 16 show that stakeholder success is greater with greater flexibility to change. We will assume that this will add to a customer's assessment of a project's success. This suggests that Agile processes, which embrace change and adaptability, can be advantageous for customers in gaining a competitive edge.

One can ask, however: What about projects that don't have a goal of increasing competitive advantage? Projects can have other goals such as cost savings, meeting regulatory requirements, maintenance and repairs, etc. It might be advisable to reword this principle in a more general manner.

P2: We can therefore confirm this principle. However, we can also note that V4 and P2 are very similar and perhaps the same thing. Should they be separate?

3.5.3 P3 - Deliver working software frequently, from a couple of weeks to a couple of months, with a preference to the shorter timescale.

For this principle we examined whether projects used an iterative approach.

Table 22 ANOVA Success and Iterative Approach: During the project, how often was an iterative approach taken?

Iterative approach	Means and ANOVA				
	Success stakeholder	Success overall	Success budget/time	Success scope/quality	Valid N
Never	0.638	0.638	0.593	0.682	39
Mostly not	0.625	0.630	0.600	0.669	113
Most times	0.704	0.701	0.657	0.738	193
Always	0.765	0.742	0.683	0.756	97
Total	0.691	0.686	0.642	0.720	442
Sig.	**0.000**	**0.000**	**0.022**	**0.001**	

Iterative approaches are clearly associated with better success. Particularly in less Agile projects. Iterative approaches clearly have a benefit to projects that are not fully Agile. There is no significance for more Agile projects, but this is likely because there are almost no Agile projects that don't use an iterative approach.

We then went on to explore what frequency of delivery was associated with the most success.

Table 23 ANOVA Sprint Length and Success (sprint length <= 15 days)

	Means and ANOVA

https://doi.org/10.52202/075831-0001

Success and sprint length	Success budget/time	Success stakeholder	Success Scope/quality	Valid N
1	0.675	0.763	0.740	9
2	0.622	0.733	0.765	7
3	0.625	0.717	0.686	8
4	0.583	0.722	0.779	6
5	0.698	0.748	0.738	27
6	0.786	0.733	0.657	2
7	0.802	0.831	0.805	13
9	0.786	1.000	0.929	1
10	0.631	0.670	0.692	116
12	0.464	0.567	0.621	2
14	0.652	0.769	0.788	15
15	0.654	0.695	0.740	33
Total	0.653	0.704	0.721	239
p(F)	0.490	0.146	0.475	

Table 24 ANOVA Sprint Length and Success (sprint length >= 15 days)

Success and sprint length	Means and ANOVA			
	Success budget/time	Success stakeholder	Success Scope/quality	Valid N
18	0.643	0.733	0.900	1
20	0.636	0.703	0.770	31
21	0.536	0.753	0.716	10
22	0.554	0.567	0.611	4
24	0.786	0.933	0.929	1
25	0.750	0.733	0.843	2
28	0.786	0.867	0.829	1
30	0.610	0.644	0.695	15
40	0.381	0.667	0.667	3
45	0.321	0.567	0.621	2
60	0.411	0.717	0.654	4
90	0.393	0.467	0.486	2
100	0.976	0.889	0.943	3
140	0.643	0.800	0.929	1
160	0.429	0.600	0.586	1
365	0.643	0.400	0.486	1
Total	0.598	0.691	0.730	82
p(F)	0.062	0.295	0.098	

The p value does not indicate a statistical relationship: there are no indications of a relationship between sprint length and project success. It may be that the best sprint length is project dependent, and teams are choosing the right sprint length for their project. A casual glance at the table points to 4-week (20 workdays) sprint lengths as being associated with the highest rates of success: for overall success, budget and time.

 https://doi.org/10.52202/075831-0001

We can confirm that using iterative approaches that deliver software frequently is associated with better success. However, the length of sprints does not seem to have a clear impact on project success. So, this principle could be rewritten.

P3: We can therefore confirm that using iterative approaches that deliver software frequently is associated with better success.

3.5.4 P4 - Business people and developers must work together daily throughout the project.

To examine this principal, we examined how often product owners or business sponsors met with a team.

Table 25 ANOVA Frequency of Meetings with Product Owners or Business Sponsor and Success (All Groups)

Frequency of meetings with product owners or business sponsor	Means and ANOVA				
	Success stakeholder	Success Overall	Success budget/time	Success scope/quality	Valid N
Every 6 months or less	0.630	0.621	0.524	0.700	18
Every 2 months	0.595	0.621	0.615	0.680	26
Monthly	0.650	0.658	0.636	0.696	98
Weekly	0.718	0.706	0.653	0.736	247
Daily	0.706	0.695	0.659	0.709	59
Total	0.691	0.686	0.642	0.719	448
p(F)	**0.002**	**0.018**	0.184	0.269	

The frequency of meetings with product owners and sponsors has a clear correlation with stakeholder success, as shown in Table 25. However, it does not have a significant impact on budget/time or scope/quality success for Agile projects compared with less Agile projects. This lack of impact may be attributed to the fact that Agile projects already have regular contact with product owners and sponsors built into their design. Interestingly, projects with weekly meetings reported the highest success rates compared to those with daily meetings. Therefore, we can conclude that the current principle may not be entirely accurate as stated.

P4: Not confirmed. Weekly meetings appear to be more effective, though regular meetings are important.

3.5.5 P5 - Build projects around motivated individuals. Give them the environment and support they need and trust them to get the job done.

Examining this principle can be challenging, as defining motivation can be subjective. It raises questions about whether we are referring to motivation towards the project, the company, or the team. However, it is common sense that an unmotivated team is unlikely to yield good results.

To explore this principle in terms of motivation, environment, and support, we focused on teams that were fully dedicated to a project.

Table 26 ANOVA Team Dedication Level to Project and Success (All Groups)

Team Fully dedicated to project	Means and ANOVA				
	Success stakeholder	Success Overall	Success budget/time	Success scope/quality	Valid N
Very Low	0.631	0.645	0.676	0.644	13
Low	0.560	0.563	0.530	0.602	48
Medium	0.647	0.656	0.626	0.705	120
High	0.731	0.717	0.661	0.744	183
Very High	0.752	0.738	0.685	0.764	84
Total	0.691	0.686	0.642	0.719	448
p(F)	**0.000**	**0.000**	**0.002**	**0.000**	

The results from Table 26 indicate that team dedication is positively correlated with all aspects of project success across different types of projects. However, it is important to note that many organizations face multiple demands and may not be able to fully dedicate teams, which can put project success at risk. In the case of more Agile projects, partial team dedication is notably linked to failed projects.

Therefore, while we cannot confirm the principle as written from a quantitative standpoint is evident that teams that are dedicated and focused on a project are associated with greater project success.

P5: We cannot confirm the principle as written from a quantitative standpoint. But teams that are dedicated and focused on a project are associated with greater project success.

3.5.6 P6: The most efficient and effective method of conveying information to and within a development team is face-to-face conversation.

To investigate this principle, we analyzed the correlation between team co-location and project success. Initially, we conducted a regression analysis. As shown in Table 27, The R^2 value (0.002) indicate a very weak association between team colocation and project success. Additionally, the p-value of 0.345 suggests that this relationship is not statistically significant.

Table 27 Colocation and Success

Colocation of team	R^2	p
Success Overall	0.002	0. 345
Valid N	449	

Further analysis was conducted using ANOVA to compare success measures across different levels of team co-location, as shown in Table 28. The results demonstrate that collocation is not significant for success. The flies against received wisdom about Agile. However, this may

https://doi.org/10.52202/075831-0001

be due to the effect that instant communication is getting better and that the structure and daily meetings that are part of Agile processes help distributed teams be more successful.

Table 28 ANOVA Team Collocation and Success (All Groups)

Local vs. remote team	Means and ANOVA				
	Success stakeholder	Success overall	Success budget/time	Success scope/quality	Valid N
Distributed internationally	0.678	0.684	0.651	0.730	129
Distributed nationally	0.712	0.705	0.646	0.747	86
Distributed in one city or region	0.678	0.671	0.644	0.685	74
Collocated in same office	0.694	0.683	0.634	0.708	157
Total	0.690	0.685	0.643	0.718	446
Sig.	0.586	0.675	0.930	0.134	

To further investigate this relationship, we analyzed data from Dataset 2, as shown in Table 29. Teams that can clearly benefit from Agile if the teams are international (located in more than one country) or national (located in one country but in multiple cities). Teams that are collocated (one city or region) do not seem to gain that clear benefit of Agile possible because communication is already good. Therefore, P6 is false. Teams do not have to be collocated to benefit from Agile or deliver successful projects.

Table 29 The impact of Agile on success by location of team from dataset 2

Location of Team Members	**Number of Projects**	**R^2**	**p-value**
International	350	.309	0.040
National	539	.257	0.002
One city or region	116	.418	0.134

P6: This was found to be false as written from a quantitative standpoint. Teams do not appear to need face-to-face communication for successful projects.

3.5.7 P7 - Working software is the primary measure of progress.

This principle is closely related to Value 2 (Working software over comprehensive documentation). In examining Value 2, we found that documentation is also important for project success. Specifically, the presence of a Work Breakdown Structure (WBS) was associated with improved success, as shown in Table 5. Additionally, the absence of a set budget was linked to less successful projects in terms of time and budget goals, as indicated in Table 19. This need for a comprehensive approach that considers both working software and documentation may explain why most projects delivered today follow a hybrid approach than pure Agile (Gemino et al., 2021).

Therefore, while this principle is valuable for many projects, we cannot fully confirm it. It may be more appropriate to restate it as an important measure of progress that should be used in conjunction with other measures for a more holistic evaluation. For example, if a project has a hardware component, just measuring the software progress is not adequate.

https://doi.org/10.52202/075831-0001

In conclusion, Principle 7 cannot be fully confirmed based on our analysis. It could perhaps be restated that is as an important measure of progress but may need to be used with other measures for many projects.

P7: We cannot fully confirm this principle. It could perhaps be restated that is as an important measure of progress but may need to be used with other measures for many projects.

3.5.8 P8 - Agile processes promote sustainable development. The sponsors, developers, and users should be able to maintain a constant pace indefinitely.

Assessing the validity of this principle from a quantitative standpoint is challenging. The principle's intention may be unclear as it deals with the notion of indefinite sustainability, which is difficult to measure in the context of project development. Agile projects, like any other projects, have defined lifecycles and will eventually reach a conclusion. It is not feasible for teams to maintain a constant pace indefinitely.

It is possible that this principle is meant to emphasize the importance of reducing pauses or delays in a development project, ensuring a steady and efficient workflow. However, without further clarification, it is challenging to evaluate this principle based on the available quantitative data. In conclusion, due to the constraints of project lifecycles and the lack of clarity regarding the purpose of this principle, we cannot examine Principle 8 as written from a quantitative standpoint.

P8: We cannot examine this principle as written from a quantitative standpoint.

3.5.9 P9 - Continuous attention to technical excellence and good design enhances agility.

While assessing technical excellence and good design can be subjective and time-dependent, it is difficult to argue against the importance of these aspects. It is evident that technical excellence and good design are desirable attributes for any project. Who would advocate for technical mediocrity and poor design? This aligns with common sense and best practices in the field. These goals are essential for achieving high-quality and successful outcomes in both Agile and non-Agile projects. Therefore, based on the available information:

P9: We cannot disprove the principle and appears evidently correct.

3.5.10 P10: Simplicity--the art of maximizing the amount of work not done--is essential.

While it is difficult to argue against the value of working efficiently, it is crucial to critically examine the principle in terms of its goal and application. Simplicity is indeed a standard goal in business, as it can lead to streamlined processes and increased productivity. However, it is essential to acknowledge that leading-edge technologies or complex projects may inherently involve a level of intricacy that cannot be simplified without compromising the desired outcomes. Oversimplification can hinder progress and overlook critical details, leading to suboptimal results. It is crucial to strike a balance between simplicity and the necessary complexity to achieve project objectives effectively.

https://doi.org/10.52202/075831-0001

While we cannot disprove the principle from a quantitative standpoint, it is important to consider the context in which it is applied. The principle may need further clarification and a more nuanced understanding to ensure its practicality and effectiveness in different scenarios.

P10: We cannot disprove the principle from a quantitative standpoint. But the statement is questionable in goal and application.

3.5.11 P11- The best architectures, requirements, and designs emerge from self-organizing teams.

The analysis of Dataset 1 provides insights into the correlation between self-organizing teams and project success. Table 30 shows the regression analysis results for self-organizing teams and overall success, with an R^2 value of 0.082 and a statistically significant p-value of 0.000. These findings indicate that self-organizing teams have a positive association with overall project success.

Table 30 Regression Self-organizing teams and Success

Self-Organizing and Success	R^2	p
Success Overall	0.082	0.000
Valid N		449

However, it is difficult to measure whether the best architectures and designs were created by these teams. Our data does allow us to check the quality of requirements. From the table below, we can see that self-organizing teams are associated with better requirements quality.

Table 31 Regression Self-organizing and Success

Self-Organizing and Requirements Quality	R^2	p
Functional Requirements	0.009	0.023
Non-Functional Requirements	0.021	0.001
Valid N		449

It is not clear from this principle is stating that team management is not required for self-organizing teams. For this question, we look at Hybrid Agile projects. A Hybrid project will usually have aspects of waterfall methodologies such as a project plan, a budget and a project manager. Many Agile teams also include a project manager or other management other than the self-organizing team. However, for this question, we will compare Hybrid processes which usually have managers involved.

Table 32 ANOVA Methodology Type and Success (All Groups)

Percentage Agile/iterative	Means and ANOVA				
	Stakeholder success	Success overall	Success budget/time	Success scope/quality	Valid N
0%	0.664	0.662	0.617	0.703	98

1-19%	0.604	0.630	0.646	0.665	47
20-39%	0.660	0.680	0.663	0.735	53
40-59%	0.694	0.687	0.645	0.713	103
60-79%	0.728	0.709	0.631	0.748	81
80-100%	0.764	0.736	0.674	0.741	67
All groups	0.691	0.686	0.643	0.719	449
P(F)	**0.000**	**0.020**	0.676	0.144	

It appears from these results (Table 32) that Agile projects are more successful and the more Agile the project, the more successful. The exception is the budget time success and, to a lesser extent, the scope quality. For those success measures, hybrid projects appear to be as successful or more successful than fully Agile projects. Gemino et al. (2021) using a sophisticated analysis found that Hybrid projects are as successful as Agile projects. An analysis of dataset two found the same thing. For efficiency (time/budget/scope) measures, Hybrid methods were as successful as Agile methods.

P11: There is evidence that self-organizing teams are associated with more project success. There is less evidence that other team management is not still required.

3.5.12 P12 - At regular intervals, the team reflects on how to become more effective, then tunes and adjusts its behaviour accordingly.

This would seem to be a rather uncontroversial principle as regular sprint reviews and adjusts should improve performance. In the table below, we did a regression against overall success and found that to be the case.

Table 33 Sprint Review Use and Success

Sprint Review Use and Success	R^2	p
Success Overall	0.011	0.014
Valid N		449

However, the R^2 is very low which may indicate a borderline effect. We did some further analysis and found the following for time and budget success measures.

Table 34 Sprint Review Use and Success

Sprint Review Use and Success	R^2	p	Coefficient
Success Budget	0.003	0.125	-0.262
Success Time	0.001	0.228	0.216
Valid N		449	

This analysis indicates that sprint reviews do not improve time success measure as we see from p not being significant at .228. In addition, for budget success measures, it is also not significant but is closer to significance with a negative coefficient. This may imply that sprint reviews do not help with on time delivery and may be a negative in meeting budget goals. Though the

https://doi.org/10.52202/075831-0001

budget result is not at the $p<.05$ significance. One could interpret this as indicating sprint reviews can be a poor use of time and money rather than a net benefit.
P12: There is some evidence that this principal can have a marginal impact on project success but also evidence that its impacts are negative for time and budget goals in projects.

4. Conclusions

Table 35 provides an overview of the confirmation status for each value and principle in Agile methodology based on the analysis conducted. The confirmation status indicates whether the value or principle is confirmed, partially confirmed, not confirmed, false, or cannot be confirmed. The notes column provides additional insights and explanations for each confirmation status.

Table 35 Confirmation of Values and Principles in Agile Methodology

Value/ Principal	Confirmation	Notes	Should be reworded?
V1	Not Confirmed	This statement is hard to verify and hard to refute.	Yes
V2	False	The opposite appears to be true in some cases, better documentation associated with successful Agile projects.	Yes
V3	Confirmed	Customer collaboration is shown to be important for project success	No
V4	Confirmed	Openness to change is associated with project success with the caveat of having budget flexibility is not.	No
P1	Confirmed	Agile processes lead to higher stakeholder success	Yes
P2	Confirmed	Flexibility to changing requirements is associated with more successful projects.	Yes
P3	Partially Confirmed	We can therefore confirm that using iterative approaches that deliver software frequently is associated with better success. However, the length of sprints does not seem to have a clear impact on project success.	Yes
P4	False	Weekly meetings appear to be more effective than daily meetings, though regular meetings are important.	Yes
P5	Cannot be confirmed	Principal is too general but teams that are dedicated and focused on a project are associated with greater project success.	Yes
P6	False	Co-location is not associated with greater success. Distribute project teams seem to have similar success.	Yes
P7	Cannot be confirmed	Could perhaps be restated that is as an important measure of progress	Yes
P8	Cannot be confirmed	We cannot examine this principle as written from a quantitative standpoint. Should be rewritten	Yes
P9	Confirmed	We cannot disprove the principle and appears evidently correct.	Yes

P10	Cannot be confirmed	We cannot disprove the principle from a quantitative standpoint. But the statement is questionable in goal and wording.	Yes
P11	Confirmed	Self-organizing teams are associated with more project success. There is less evidence that other team management is not still required.	No
P12	Cannot be confirmed	This principal can have a marginal impact on project success but also evidence that its impacts are negative for time and budget goals	Yes

Therefore, out of the 16 hypotheses, 6 were confirmed as clearly useful, 7 were partially correct or could not be confirmed either way and 3 were found to be false. We find that the majority would benefit from rewording.

In this comprehensive examination of Agile values and principles, we find a varied landscape of confirmations, refutations, and nuanced insights. The Agile values, particularly V1 and V2, present challenges in quantifiable verification. The emphasis on "individuals and interactions over processes and tools" (V1) lacks clear evidence either confirming or refuting its association with project success. Similarly, the assertion that "working software over comprehensive documentation" (V2) is a key driver for success is contradicted by findings suggesting the importance of certain documentation, such as non-functional requirements and work breakdown structures.

On the other hand, some of the principles behind the Agile Manifesto offer more concrete insights. Customer collaboration (V3) emerges as a confirmed positive factor, with engaged stakeholders significantly contributing to project success. The openness to change (V4) is confirmed, provided there is caution with budget flexibility. The principles associated with Agile processes, such as early and continuous delivery (P1) and adapting to changing requirements (P2), exhibit positive correlations with stakeholder satisfaction and project success.

However, some principles require nuanced considerations. The notion that "delivering working software frequently" (P3) is associated with success is affirmed, but the impact of sprint length on success remains unclear. The emphasis on daily collaboration (P4) is challenged, suggesting that weekly meetings may be more effective in certain contexts. The ideal of "building projects around motivated individuals" (P5) lacks direct confirmation, but team dedication emerges as a crucial factor in project success. Additionally, the principles related to team co-location (P6) and the primary measure of progress (P7) encounter challenges in full confirmation. The sustainability principle (P8) is difficult to examine quantitatively, and the call for continuous attention to technical excellence (P9) is acknowledged as a positive goal without definitive quantitative confirmation.

While the principles of simplicity (P10) and self-organizing teams (P11) are not disproven, their application may be context-dependent. Finally, the principle of regular reflection (P12) shows a marginal impact on project success, but caution is advised as it may have negative repercussions on time and budget goals in certain scenarios. In conclusion, this analysis underscores the nuanced and context-dependent nature of Agile values and principles.

https://doi.org/10.52202/075831-0001

5. Recommendations

The Agile manifesto stands as a pivotal document in the realm of software development, serving as a guiding principle for numerous projects over the past two decades. However, its longevity prompts reflection on its continued relevance and applicability in contemporary contexts. It is evident that the landscape has evolved, necessitating a critical assessment of its tenets.

Numerous voices within both academic and non-academic circles have advocated for a review and potential revision of the Agile Manifesto. Hohl et al. (2018) discuss the divergence of opinions among the original authors regarding the need for revision, with some advocating for updates while most advocate maintaining its original form. Additionally, Kent Beck's "Beyond Agile Manifesto" initiative in 2011 underscores a recognition of the need for reevaluation, albeit with limited traction.

While the Agile Manifesto does not dictate specific methodologies or project management practices, its influence on project execution is undeniable. Thus, any proposed revisions should be approached with careful consideration of their potential impact.

This study aligns with the notion that the Agile manifesto warrants a review, although perhaps not to the extent proposed by Beck et al (2001) . Given the advancements and insights gained over two decades, an update, albeit moderate, may be warranted, potentially timed to coincide with the manifesto's 25th anniversary. Such an update would not only acknowledge the evolution of Agile and project management practices but also ensure its continued relevance for the next generation of projects.

https://doi.org/10.52202/075831-0001

References

Beck, K., Beedle, M., Van Bennekum, A., Cockburn, A., Cunningham, W., Fowler, M., Grenning, J., Highsmith, J., Hunt, A., & Jeffries, R. (2001). *Manifesto for agile software development*. Agile Alliance. https://agilemanifesto.org/

Behutiye, W., Rodriguez, P., & Oivo, M. (2022). Quality requirement documentation guidelines for agile software development. *IEEE Access, 10*, 70154–70173. https://doi.org/10.1109/ACCESS.2022.3187106

Bose, I. (2008). Lessons learned from distributed agile software projects: A case-based analysis. *Communications of the Association for Information Systems, 23*. https://doi.org/10.17705/1CAIS.02334

Cockburn, A., & Highsmith, J. (2001). Agile software development, the people factor. *Computer, 34*(11), 131–133. https://doi.org/10.1109/2.963450

Crawford, L., Turner, J. R., & Hobbs, B. J. (2004). Project categorization systems and their use in organisations: An empirical study. *Project Management Institute Research Conference*.

Deneckere, R., Kornyshova, E., & Iacovelli, A. (2016). Progressive integration of method components: A case of agile IS development methods. *2016 IEEE Tenth International Conference on Research Challenges in Information Science (RCIS)*, 1–10. https://doi.org/10.1109/RCIS.2016.7549347

Denning, S. (2015). Agile: It's time to put it to use to manage business complexity. *Strategy & Leadership, 43*(5), 10–17. https://doi.org/10.1108/SL-07-2015-0057

Dingsøyr, T., Nerur, S., Balijepally, V., & Moe, N. B. (2012). A decade of agile methodologies: Towards explaining agile software development. *Journal of Systems and Software, 85*(6), 1213–1221. https://doi.org/10.1016/j.jss.2012.02.033

Fitsilis, P., Gerogiannis, V., & Anthopoulos, L. (2014). Ontologies for software project management: A Review. *Journal of Software Engineering and Applications, 07*(13), 1096–1110. https://doi.org/10.4236/jsea.2014.713097

Gemino, A., Horner Reich, B., & Serrador, P. M. (2021). Agile, traditional, and hybrid approaches to project success: Is hybrid a poor second choice? *Project Management Journal, 52*(2), 161–175.

Hohl, P., Klünder, J., Van Bennekum, A., Lockard, R., Gifford, J., Münch, J., Stupperich, M., & Schneider, K. (2018). Back to the future: Origins and directions of the "Agile Manifesto" – views of the originators. Journal of Software Engineering Research and Development, 6(1), 15. https://doi.org/10.1186/s40411-018-0059-z

Kadenic, M. D., Koumaditis, K., & Junker-Jensen, L. (2023). Mastering scrum with a focus on team maturity and key components of scrum. Information and Software Technology, 153, 107079. https://doi.org/10.1016/j.infsof.2022.107079

Kantola, K., Vanhanen, J., & Tolvanen, J. (2022). Mind the product owner: An action research project into agile release planning. Information and Software Technology, 147, 106900. https://doi.org/10.1016/j.infsof.2022.106900

Kinnunen, H., & Luoma, E. (2018). Towards measuring the agility of software business. *Hawaii International Conference on System Sciences*. https://doi.org/10.24251/HICSS.2018.677

Kuhrmann, M., Tell, P., Hebig, R., Klünder, J., Münch, J., Linssen, O., Pfahl, D., Felderer, M., Prause, C. R., MacDonell, S. G., Nakatumba-Nabende, J., Raffo, D., Beecham, S., Tüzün, E., López, G., Páez, N., Fontdevila, D., Licorish, S. A., Kupper, S., … Richardson, I. (2022). What makes agile software development agile? *IEEE Transactions on Software Engineering, 48*(9), 3523–3539. https://doi.org/10.1109/TSE.2021.3099532

 https://doi.org/10.52202/075831-0001

Looks, H., Fangmann, J., Thomaschewski, J., Escalona, M.-J., & Schön, E.-M. (2021). Towards a standardized questionnaire for measuring agility at team level. In P. Gregory, C. Lassenius, X. Wang, & P. Kruchten (Eds.), *Agile processes in software engineering and extreme programming* (Vol. 419, pp. 71–85). Springer International Publishing. https://doi.org/10.1007/978-3-030-78098-2_5

Luna, A., Costa, C., Moura, H., Novaes, M., & Nascimento, C. (2010). Agile governance in information and communication technologies: Shifting paradigms. *JISTEM Journal of Information Systems and Technology Management, 7*(2), 311–334. https://doi.org/10.4301/S1807-17752010000200004

Madi, T., Dahalin, Z., & Baharom, F. (2011). Content analysis on agile values: A perception from software practitioners. *2011 Malaysian Conference in Software Engineering*, 423–428. https://doi.org/10.1109/MySEC.2011.6140710

Müller, R., & Turner, J. R. (2007). Matching the project manager's leadership style to project type. *International Journal of Project Management, 25*(1), 21–32.

Rahy, S., & Bass, J. M. (2022). Managing non-functional requirements in agile software development. *IET Software, 16*(1), 60–72. https://doi.org/10.1049/sfw2.12037

Salinas, M. R. N., Neto, A. G. S. S., & Emer, M. C. F. P. (2018). Concerns and limitations in agile software development: A survey with Paraguayan companies. In V. A. D. Santos, G. H. L. Pinto, & A. G. Serra Seca Neto (Eds.), *Agile methods* (Vol. 802, pp. 77–87). Springer International Publishing. https://doi.org/10.1007/978-3-319-73673-0_6

Sandstø, R., & Reme-Ness, C. (2021). Agile practices and impacts on project success. *Journal of Engineering, Project, and Production Management, 11*(3), 255–262.

Schmitt, A., & Hörner, S. (2021). Systematic literature review – improving business processes by implementing agile. Business Process Management Journal, 27(3), 868–882. https://doi.org/10.1108/BPMJ-10-2019-0422

Serrador, P., & Pinto, J. K. (2015). Does Agile work?—A quantitative analysis of agile project success. International Journal of Project Management, 33(5), 1040–1051.

Serrador, P., & Turner, R. (2015). What is enough planning? Results from a global quantitative study. IEEE Transactions on Engineering Management, 62(4), 462–474.

Shenhar, A. J., Dvir, D., Levy, O., & Maltz, A. C. (2001). Project success: A multidimensional strategic concept. Long Range Planning, 34(6), 699–725.

Tolfo, C., Wazlawick, R. S., Ferreira, M. G. G., & Forcellini, F. A. (2018). Agile practices and the promotion of entrepreneurial skills in software development. Journal of Software: Evolution and Process, 30(9), e1945. https://doi.org/10.1002/smr.1945

Vierlboeck, M., Gövert, K., Trauer, J., & Lindemann, U. (2019). Budgeting for agile product development. *Proceedings of the Design Society: International Conference on Engineering Design, 1*(1), 2169–2178. https://doi.org/10.1017/dsi.2019.223

Biography

Pedro M. Serrador, PMP, ACP, P.Eng., MBA, PhD is a writer and researcher on Agile, AI and management topics. He is an adjunct professor at the Northeastern University and University of Toronto, Canada. He is also the owner of Serrador Project Management, a consultancy in Toronto, Canada. He has worked on projects in the financial, telecommunications, utility, medical imaging, and simulations sectors for some of the world's largest companies.

He an author of books and articles on project management and is also a regular speaker at conferences. He was the recipient of the PMI 2012 James R. Snyder International Student Paper of the Year Award, the Major de Promotion Award for best PhD Thesis 2012-2013 from SKEMA business school and the 2022 Most cited paper of the year award from Project Management Journal. He holds a Hons. BSc in Physics and Computer Science from the University of Waterloo, Canada; an MBA from Heriot-Watt University, Edinburgh, Scotland; and a PhD in Strategy, Programme & Project Management from SKEMA Business School (Ecole Supérieure de Commerce de Lille, France).

https://doi.org/10.52202/075831-0001

Strategic Value-Driven Medical Affairs: The Leadership Role of Project and Program Management

Premanand Sundivakkam, PhD., MBA., BCMAS
Bratati Ganguly, PhD., MBA., BCMASc
Sriramasundararajan Rajagopalan, PhD., MBA, PMP, PfMP, PgMP
Shashi Singh MBA., PMP., CSP., SAFe, SPC4

Abstract

This article explores the crucial role of Medical Affairs in the pharmaceutical and medical devices industries, particularly within the rapidly evolving digital landscape. It highlights the fundamental objectives of Medical Affairs, with a specific focus on the concept of "value." The article underscores the necessity of strategic alignment and value optimization as key drivers in enhancing the value that Medical Affairs brings to these industries and, ultimately, to patient care. Offering a current overview of practices within the Medical Affairs function, the article spotlights the diversity of approaches and the emerging demands triggered by the digital age and generational shift. It identifies challenges and untapped opportunities in optimizing the value of Medical Affairs, emphasizing the pressing need for change. Additionally, the article addresses potential challenges in the evolution of Medical Affairs and concludes with a future-oriented perspective on its role. It asserts that strategic alignment and value optimization are not merely desirable but essential in the realm of project management within Medical Affairs. When project managers demonstrate leadership by connecting their project delivery skills to the benefits management lifecycle promoted by program management, such knowledge becomes indispensable for Medical Affairs. By embracing these integrated disciplines, Medical Affairs can better serve patients and address healthcare needs in an ever-changing landscape, thereby delivering significant value to the pharmaceutical and medical devices industries. Supported by relevant references, this article provides a comprehensive and insightful read for anyone interested in strategic alignment and value optimization in Medical Affairs.

Keywords: medical affairs, value optimization, patient care, leadership, project management, program management, benefits management, risk management

https://doi.org/10.52202/075831-0002

Introduction

Medical Affairs has emerged as a critical function in the pharmaceutical and medical devices industries. It serves as the third strategic pillar alongside Research & Development (R&D) and commercial operations (Evers et al., 2019; Algazy et al., 2023).

Medical Affairs has transformed to collaborate effectively with R&D and commercial operations, propelled by the necessity for strategic alignment and value optimization. This transformation is gaining significance considering swift progress in digital technology, biological sciences, and data generation (Evers et al., 2019; Algazy et al., 2023). The article presents a contemporary overview of practices within the Medical Affairs function, highlighting the variety of approaches and the emerging requirements sparked by these advancements.

At the core of Medical Affairs lies scientific communication, which is now strategically aligned with the overarching organizational objectives and optimized for maximum value (A MAPS White Paper, 2020). This involves fostering information exchange about products with healthcare professionals (HCP), academics, and regulatory bodies, ensuring these stakeholders have precise and current information (A MAPS White Paper, 2020). This information exchange is not about creating reports or facilitating meetings but about leading collaboration with multiple stakeholder groups unearthing individual, interdependent, and overall risks impacting value. Consequently, this strategic alignment and value optimization in communication directly influence patient care and outcomes, underlining the urgent need for transformation[3].

In the realm of medical devices, Medical Affairs provides unique clinical expertise, professional education, and medical safety (Pracyk et al., 2023). These functions are essential for ensuring that medical devices are used safely and effectively, thereby enhancing patient care. The ubiquitous nature of technology in the healthcare industry has also created innovative Software as a Medical Device (SaMD) besides Software in Medical Devices (SiMD) further emphasizes (Gerke et al., 2020) the convergence of multiple domain expertise connected by strategic project and program management.

The article also discusses potential hurdles in the evolution of Medical Affairs (Algazy et al., 2023) and concludes with a forward-looking viewpoint on its role. It asserts that strategic alignment and value optimization are not merely preferable but crucial in the domain of Project Management within Medical Affairs. When project managers exhibit leadership by linking their project delivery skills to the benefits management lifecycle advocated by program management, this wealth of knowledge becomes invaluable for Medical Affairs (CMK Select Insights, 2023).

By adopting these integrated disciplines, Medical Affairs can more effectively serve patients and address healthcare needs in a constantly changing landscape, thereby delivering substantial value to the pharmaceutical and medical devices industries (Evers et al., 2019; Algazy et al., 2023). This article offers a comprehensive and insightful exploration of strategic

https://doi.org/10.52202/075831-0002

alignment and value optimization in Medical Affairs, making it an essential read for a wider audience (Algazy et al., 2023; Evers et al., 2019).

Understanding "Value" in the Context of Medical Affairs

Medical Affairs stands out as a crucial component of healthcare, drawing its strength from a unique combination of scientific and medical knowledge (Shepard et al., 2024). This value is demonstrated through the generation and dissemination of high-quality, evidence-based medical information that supports the safe and effective use of medical products (Shepard et al., 2024) The primary focus of Medical Affairs is on generating real-world evidence (RWE), nurturing peer-to-peer connections, and providing educational opportunities for healthcare professionals and stakeholders[7]. These efforts collectively work towards enhancing patient outcomes, highlighting the pivotal role of Medical Affairs in fostering innovation and improving patient care (Aherne et al., 2022; John, 2023; Ramadas et al., 2022).

The value proposition of Medical Affairs in patient care is unequivocal. It ensures the availability of essential medications to those in need, addressing key issues like awareness, access, and treatment adherence. By concentrating on these critical areas, Medical Affairs significantly contributes to enhancing patient outcomes and their overall well-being (John et al., 2024).

To fully leverage the potential of Medical Affairs in improving patient outcomes, enhancing treatment strategies, and strengthening trust and credibility within the medical community and beyond, it is crucial that the organizations adopt efficient strategies that prioritizes strategic alignment and value maximization (Algazy et al., 2023).

The Imperatives of Strategic Alignment for Enhancing Medical Affairs Value

Strategic alignment within Medical Affairs is crucial for maximizing its value and impact across the pharmaceutical and medical devices industries. This alignment involves synchronizing Medical Affairs objectives with the broader organizational goals, including but not limited to R&D, commercial, regulatory, and market access strategies. By ensuring that Medical Affairs activities are integrated and supportive of the company's strategic vision, the department can more effectively contribute to product development, regulatory success, and market penetration.

Strategic alignment enables Medical Affairs to provide timely and relevant scientific insights that inform clinical trial designs, support regulatory submissions, and shape evidence generation efforts that meet both clinical and market needs. Additionally, it facilitates cohesive communication and collaboration with key stakeholders, including HCPs, key opinion leaders (KOLs), and patient advocacy groups, thereby enhancing the credibility and acceptance of the

https://doi.org/10.52202/075831-0002

company's products (Algazy et al., 2023). Ultimately, a strategically aligned Medical Affairs function enhances its value by driving innovation, ensuring regulatory compliance, and improving patient outcomes through well-informed, patient-centric approaches. This not only enhances patient care but also ensures efficient resource utilization, ultimately amplifying the value that Medical Affairs brings to the organization.

Transforming Medical Affairs: Embracing Digitalization, RWE, and Patient-Centricity

Medical Affairs is dynamically evolving, integrating digital technologies, data analytics, and patient-centric approaches. The use of digital platforms for engagement and communication has increased, particularly during the COVID-19 pandemic (Ramadas et al., 2020). RWE generation is now a key focus, aiding in understanding the real-world impact of therapies. Cross-functional collaboration with R&D, commercial, regulatory, and market access teams is emphasized to align with business objectives (MAPS 2020 Report). Engagement with KOLs and patient advocacy groups is being strengthened to centralize patient insights in therapeutic development. Advanced data analytics are being adopted to derive actionable insights from clinical and real-world data, supporting evidence-based decision-making and strategic planning (Ramadas et al., 2020; MAPS 2020 Report).

The approach to Medical Affairs varies significantly between large and small companies, each carving its unique path influenced by resources, organizational structures, and strategic priorities. Despite their differences, both large and small pharmaceutical companies share a common goal: ensuring the safe, effective, and evidence-based use of their products.

Large Pharmaceutical Companies: Depth and Breadth – Large pharmaceutical companies are often characterized by extensive Medical Affairs departments. These departments are composed of specialized teams dedicated to various therapeutic areas and functions such as medical communications, health economics, and outcomes research (HEOR), and RWE generation[1]. With substantial budgets at their disposal, these companies invest heavily in cutting-edge technologies, large-scale clinical trials, and global KOL engagements (Rickwood et al., 2022; Evers et al., 2014). This investment allows for a broad and deep scientific exchange, enhancing their ability to deliver comprehensive healthcare solutions.

Small Pharmaceutical Companies: Agility and Innovation – On the other hand, small pharmaceutical companies or biotech firms often operate with leaner teams. In these settings, Medical Affairs professionals may wear multiple hats (de Bono, 1999; Rajagopalan, 2016), handling a broader range of responsibilities. Despite their limited resources, these organizations focus on niche therapeutic areas or innovative treatments (Shepard et al., 2024), emphasizing agility and rapid decision-making (Whisman, 2023). They leverage digital tools and platforms

 https://doi.org/10.52202/075831-0002

for virtual engagements and rely heavily on external partnerships and collaborations to extend their reach and impact.

Embracing the Future: The Role of Medical Affairs in the Technological Revolution of Healthcare

Over time, the industry has invested heavily in recruiting data scientists and engineers, integrating data sets, and developing transformative technologies. This enables more precise and timely insights that inform clinical decision-making and product strategies. Digital platforms also facilitate virtual engagements, telemedicine, and e-learning, which are becoming essential for effective communication with HCPs and KOLs. It is imperative for Medical Affairs to stay ahead of the curve by embracing advanced technologies to drive performance, efficiency, and collaboration with stakeholders and ultimately improve patient care and outcomes (Algazy et al., 2023; Dwyer et al., 2020). Here are a few examples:

1. **Data Analytics:** By leveraging data analytics, Medical Affairs can help identify clinical gaps and trends (Dylla, 2024), leading to more informed decision-making and improved patient outcomes (Evers et al., 2019). For instance, a study published in the Journal of Medical Internet Research found that predictive analytics could identify patients at risk of chronic disease like diabetes up to 10 years in advance (Sheng et al., 2021).
2. **RWE:** Advanced RWE capabilities can support existing treatments and provide new types of data to optimize the treatment outcomes in diverse patient populations, leading to improved guidelines and personalized care (Evers et al., 2019; Algazy et al., 2023).
3. **Artificial Intelligence (AI):** AI, including natural language processing models, and GenAI, can revolutionize patient care by automating medical writing, producing real-time insights (Algazy et al., 2023), enabling precision medicine and personalized treatment plans. For example, Google's DeepMind AI has been used to predict patient deterioration up to 48 hours in advance, allowing for timely invention and improved patient outcomes (Lomas, 2019).
4. **Integration of Diverse Data Sets:** The ability to combine, analyze, and interpret disparate data sets can inform interactions with stakeholders and ultimately improve patient outcomes (Algazy J, et. al., 2023). A study in Nature Digital Medicine highlighted how integrated data sets led to improved prediction of cardiovascular risk (Hippisley-Cox et al., 2024).
5. **Personalized Medicine:** Technology that creates opportunities for more individualized treatment could greatly enable value-based care plans by increasing patient satisfaction, adherence to the treatment, and compliance as well as reducing redundancies and unnecessary treatments (Elace Health, 2022). A report from the Personalized Medicine Coalition showed that cancer patients receiving personalized treatments had a 50% higher survival rate over five years.

https://doi.org/10.52202/075831-0002

6. **Telemedicine and Remote Monitoring:** These technologies can expand access to specialized care, meeting the evolving needs of patients and HCPs (Dylla, 2024).

The urgency for change is clear: to enhance the value of Medical Affairs, we must adapt, evolve, and lead in this rapidly changing landscape. This is not just a logical step, but a necessary stride towards the future of healthcare.

Challenges and Missed Opportunities in Optimizing Medical Affairs Value

The current management practices in Medical Affairs are guided by a variety of methodologies (Terra, 2023). These methodologies, such as Agile, aim to provide flexibility and adaptability in response to changing priorities or stakeholder needs (Terra, 2023). However, despite these methodologies, there are several areas where the current practices could be enhanced to better meet the evolving demands of the healthcare landscape.

1. **Need for Comprehensive Planning**: In the realm of Medical Affairs, the lack of comprehensive planning can lead to significant project failures. It is commonly observed that the Medical Affairs projects are frequently initiated based on immediate needs or demands, without a thorough understanding of the broader implications. This reactive approach can lead to a host of problems, including misallocation of resources, unrealistic timelines, and ultimately, project failure. By adopting a more proactive and comprehensive approach to planning, Medical Affairs departments can significantly reduce the risk of project failure, ensure more efficient use of resources, and ultimately deliver better outcomes for patients and healthcare providers.

2. **Current Communication Challenges**: One of the prominent challenges in Medical Affairs organizations is the absence of a standardized and transparent approach. In the absence of a clear, uniform strategy, Medical Affairs teams often find themselves working in silos, leading to disjointed efforts and a lack of synergy. In a scenario where a clinical trial is being planned, the lack of open communication can result in the regulatory affairs team being uninformed about changes in trial protocols. This could lead to delays in obtaining necessary approvals, thereby pushing back the trial commencement date. By fostering efficient and transparent communication, Medical Affairs organizations can ensure that they are well-equipped to implement projects successfully, thereby driving innovation and delivering value to their stakeholders.

3. **Limited Stakeholder Engagement**: This is a recognized challenge in Medical Affairs organizations that can constrain project scope and inhibit innovation. For instance, without regular input from HCPs, teams may overlook critical aspects of patient care or fail to consider the practical implications of their decisions. Similarly, limited engagement with

https://doi.org/10.52202/075831-0002

patients can result in a lack of understanding of their needs and preferences, thereby reducing the patient-centricity of the project. Therefore, it is essential to foster an environment that encourages active and ongoing engagement with all stakeholders, thereby ensuring that projects are comprehensive, patient-centric, and innovative.

4. **Inadequate Resource Utilization**: Inadequate resource utilization, encompassing financial assets, human capital, time, and technology, can lead to wastage and reduced organizational value. Primary consequences include unnecessary expenditure, as seen when tasks that could be automated or outsourced are performed by staff. It also results in missed opportunities, such as failing to leverage data analytics for strategic decision-making and improved outcomes. Furthermore, it can cause project bottlenecks and delays, impacting the delivery of critical services. Addressing these challenges requires better resource planning and allocation, enhanced staff training and support, and the implementation of efficient processes and systems.

5. **Insufficient Overall Risk Management:** The current and future state of the role of Medical Affairs in delivering quality healthcare characterizes the VUCA (volatile, uncertain, complex, and ambiguous) world. Every engagement from inception to value delivery is loaded with numerous types of risks not limited to medical domain. The expertise of project and program managers with an overarching awareness of the risk management standards and healthcare specific regulations can adequately prepare the Medical Affairs value delivery effectively and efficiently.

To address these shortcomings, it is imperative to adopt a more strategic approach to project management in Medical Affairs. This approach should prioritize strategic alignment and value maximization, ensuring that all initiatives are directed toward the most relevant and impactful areas. By doing so, Medical Affairs can optimize the value delivered to the organization and its stakeholders, ultimately enhancing patient care and outcomes.

Project Management Leadership: A Key to Delivering Value in Medical Affairs

The need for effective project management leadership in Medical Affairs is clear. Project management leadership plays a pivotal role in driving the strategic alignment of Medical Affairs. It involves overseeing the execution of strategies, managing resources, and ensuring that projects are completed on time and within budget. By delivering strategies that improve benefit management and promoting an integrated discipline, project management leadership can help Medical Affairs serve patients and address healthcare needs in a rapidly changing healthcare ecosystem.

https://doi.org/10.52202/075831-0002

Effective project management leadership can:

1. **Facilitate Strategic Alignment:** By coordinating various projects and initiatives employing comprehensive planning and communication strategies, project management leadership (Rajagopalan, 2016) can ensure that all efforts are aligned with the strategic objectives of Medical Affairs. This alignment is crucial for enhancing the value of Medical Affairs in the healthcare sector (Kohles et al., 2023).
2. **Promote Collaboration:** Project management leadership fosters collaboration among various stakeholders, including healthcare professionals, patients, and regulatory authorities. This collaboration can lead to a more holistic understanding of patient needs and expectations, thereby improving patient care and outcomes (Bedenkoy et al., 2020).
3. **Drive Technological Advancements:** Project management leadership plays a key role in driving the adoption of advanced technologies (Rajagopalan, 2023) such as data analytics, AI, and telemedicine. By overseeing the implementation of these technologies, project management leadership can enhance the performance, efficiency, and stakeholder collaboration of Medical Affairs (Bedenkoy et al., 2020).
4. **Enhance Skills and Capabilities:** Project management leadership also involves enhancing the skills and capabilities of personnel. This includes providing training in advanced technologies, data analysis, and patient engagement strategies. By doing so, project management leadership can ensure that Medical Affairs has the necessary expertise to navigate the changing healthcare landscape (Lakata et al., 2024).

Advanced and Innovative Project Management Approaches

Project management has evolved significantly over the years, with innovative approaches being adopted to enhance strategic alignment and value optimization. These approaches not only facilitate strategic alignment but also promote collaboration, drive technological advancements, and enhance skills and capabilities.

1. **Agile Methodology:** Agile project management is a flexible and iterative approach that promotes collaboration and responsiveness to change (Behrens et al., 2021). It allows teams to deliver value faster and adapt to changing requirements (Moloto et al., 2020).
2. **Waterfall Methodology:** The waterfall model is a linear process where each task must be completed before the next can be started. This method is detailed and great for large projects with multiple stakeholders.
3. **Scrum:** Scrum involves short "sprints" used to create a project cycle. It promotes collaboration, enhances skills, and drives technological advancements.
4. **Artificial Intelligence and Machine Learning (ML):** AI and ML are being applied to project management to improve project selection, monitor progress, speed up reporting, and facilitate testing.

 https://doi.org/10.52202/075831-0002

5. **Lean Project Management (LPM):** LPM focuses on maximizing value and minimizing waste. It enhances skills and capabilities by promoting continuous improvement.

Empowering Medical Affairs: The Imperative of Novel Collaborations for Innovative Project Management

Medical Affairs in the pharma industry can adapt these advanced project management approaches to achieve success in strategic alignment and value optimization:

1. Agile and Scrum methodologies have been successfully adopted in various industries, including healthcare, to promote collaboration, enhance skills, and drive technological advancements (Lance, 2024). A few methodologies that project management leadership can adopt for Medical Affairs functions are:
    a. Promoting daily or weekly standup meetings where teams can discuss and share updates to continuously identify and address barriers. This practice also enhances collaboration and promotes a culture of transparency and continuous improvement.
    b. Sprint-based workflow, where work can be divided into small manageable units (sprints), allowing for rapid adaptation to change (Van Schooendervowert et al., 2021). This will enhance the skills as the medical affairs members learn to work within time constraints and prioritize tasks effectively.
    c. Agile and Scrum methodologies focus on iterative development, where feedback is continuously incorporated to improve outcomes. This can be particularly beneficial in Medical Affairs for insights generation and utilization, establishing patient engagement strategies, etc.

2. AI is transforming the landscape of project management in Medical Affairs (Bhide, 2024). Here is how these technologies can enhance various aspects of project management:
    a. Project Selection: AI and ML can analyze historical data and predict the success rate of potential projects. This predictive capability can guide Medical Affairs in selecting the most promising projects like the financial institutions that use similar AI models to predict the risk and return of investment profiles.
    b. Progress Monitoring: AI can provide real-time tracking of project progress, alerting the team to any deviations from the plan. This allows for timely interventions and course corrections. For instance, in manufacturing, AI is used to monitor production lines for anomalies, a concept that can be applied to project monitoring in Medical Affairs.
    c. Speeding up Reporting: AI can automate the generation of project reports, significantly reducing the time spent on this task. These reports can include key metrics, progress updates, and forecasts. In the e-commerce industry, AI is used to generate real-time sales reports, a practice that can be adopted by Medical Affairs.

https://doi.org/10.52202/075831-0002

3. LPM is a methodology that focuses on delivering value to the customer by eliminating waste (CMK Select Insights, 2023). In the context of Medical Affairs, the 'customer' is the patient, and 'value' is defined as any action or process that a patient would want to pay for. Here's how LPM can be adopted in Medical Affairs to maximize value and minimize waste:
    a. Value Stream Mapping: This involves mapping out every single action required to take a project from start to finish. By doing so, Medical Affairs can identify non-value adding activities (waste) and work on eliminating them.
    b. Maximizing Value: LPM can help Medical Affairs focus on activities that directly contribute to patient-centric outcomes. This could include improving the quality of medical advice, ensuring the relevance of research projects to current healthcare needs, and enhancing the effectiveness of patient communication channels.
    c. Continuous Improvement: LPM encourages continuous improvement through iterative cycles of planning, doing, checking, and acting (PDCA). This can lead to consistent enhancements in meeting healthcare needs and patient expectations.
    d. Just-in-Time Delivery: LPM promotes efficiency by advocating for 'Just-in-Time' delivery. This means that resources are utilized only when they are needed.
    e. Decentralizing Decision-Making: LPM decentralizes decision-making, empowering every team member to make decisions that speed up processes and improve quality. This can lead to quicker responses to patient needs and expectations.
    f. Building a Lean Culture: Adopting LPM is not just about implementing new processes; it is about building a lean culture. This involves training staff to identify and eliminate waste and fostering a patient-centric mindset.

By adapting Lean Project Management, Medical Affairs can drive value maximization and waste minimization. This approach can lead to improved patient-centric outcomes, better meet healthcare needs, exceed patient expectations, and optimize value, ultimately leading to enhanced patient care and satisfaction.

The Indispensable Role of Project Managers in Medical Affairs

Project managers are the linchpins in the intricate machinery of Medical Affairs, ensuring that projects are delivered on time, within budget, and align with the strategic goals of the organization (CMK Select Insights, 2023). Their work extends beyond individual projects, permeating the entire organization, influencing strategic decisions, shaping policies, and driving innovation.

Their unique set of project delivery skills, including empathy, negotiation, collaboration, and leadership, enable them to manage the complexity of project execution, ensuring the organization receives valuable results (CMK Select, 2023). These skills are not just about

https://doi.org/10.52202/075831-0002

managing projects; they are about driving change, delivering value, and shaping the future of healthcare.

Project management leadership in medical affairs brings a structured approach to planning by considering all project aspects, defining clear objectives, identifying risks, and developing a detailed project plan. Project managers act as a central communication hub to keep stakeholders informed, promote transparent communication, build trust, facilitate decision-making, and foster collaboration towards the common goal (CMK Select, 2023). Additionally, they leverage assets, maximize budget efficiencies, and allocate resources effectively to contribute to the project's overall success.

The role of project managers is not just about managing individual projects. They are not just facilitators; they are strategic partners (Bedenkoy et al., 2020; CMK Select, 2023; MAPS Midwest Summit, 2023), driving the development of new program and portfolio strategies that align with the product management lifecycles. Their role in the benefit management lifecycle in Medical Affairs extends far beyond traditional project management. They are data custodians, technology integrators, and stakeholder liaisons who can support Medical Affairs organizations to achieve desired outcomes and expected benefits that align with the organization's strategic goals (Bedenkoy et al., 2020).

With the multifaceted role, project managers have become inextricably linked to the success of many brands. They are deeply involved in crafting the medical "story" for a brand, researching, and detailing the product's patient journey, managing documentation, assessing risks, ensuring team-based efficiencies, interacting with a variety of stakeholders, and monitoring timelines and budgets.

In conclusion, project managers in Medical Affairs, by effectively connecting their project delivery skills to the benefit management lifecycle, demonstrate leadership that is indispensable for the department. Their ability to manage projects while ensuring the realization of benefits makes them an asset in the healthcare landscape. Recognizing their essential contribution is the first step towards leveraging their potential to its fullest. Their role is not just important, it is essential in the ever-evolving world of Medical Affairs.

Discussion

In the ever-evolving healthcare landscape, Medical Affairs stands as a crucial bridge between R&D, commercial operations, and patients. The role of project management leadership is pivotal in this context, not merely as facilitators but as strategic partners shaping program and portfolio strategies. Their expertise in managing projects and realizing benefits is invaluable.

As we look to the future, the adaptability of Medical Affairs to technological advancements and data analytics will shape its trajectory. Project management leadership will be instrumental in navigating this transformation, tackling challenges of digitalization, RWE generation, and patient-centricity.

Continual evolution and adaptation are imperative for project management leadership. By embracing innovative methodologies and technologies, and fostering a culture of continuous learning and improvement, they can ensure Medical Affairs stays at the forefront of healthcare innovation, delivering value to the pharmaceutical and medical devices industries, and most importantly, to patients.

In conclusion, project management leadership is not just important, but essential in Medical Affairs. Their leadership will determine how effectively Medical Affairs can navigate the future, drive innovation, and enhance patient care. Despite the challenges ahead, with strategic alignment, value optimization, and effective leadership, Medical Affairs is well-positioned to transform these challenges into opportunities for growth and innovation. The future of Medical Affairs is promising, with project management leadership guiding the way.

https://doi.org/10.52202/075831-0002

References

Assessing the value of project management in medical affairs. (n.d.). CMK Select. https://www.cmkselect.com/assessing-the-value-of-project-management-in-medical-affairs/

A vision for medical affairs in 2025. (2019, April). McKinsey & Company Insights. https://www.mckinsey.com/industries/life-sciences/our-insights/a-vision-for-medical-affairs-in-2025

A vision for medical affairs 2030: Five priorities for patient impact. (2023, October). McKinsey & Company Insights. https://www.mckinsey.com/industries/life-sciences/our-insights/a-vision-for-medical-affairs-2030-five-priorities-for-patient-impact

Aherne, S., Graves, R., Kahn, M., et al. (2022). Measuring medical affairs' value in real-time. Sorcero Webinar. https://www.sorcero.com/webinars/measuring-medical-affairs-value

Bedenkov, A., Rajadhyaksha, V., Beekman, M., et al. (2020). Developing medical affairs leaders who create the future. *Pharmaceutical Medicine, 34,* 301–307. https://doi.org/10.1007/s40290-020-00351-y

Behrens, A., Ofori, M., Noteboom, C., & Bishop, D. (2021). A systematic literature review: How agile is agile project management? *Issues in Information Systems, 22,* 278–295. https://doi.org/10.48009/3_iis_2021_298-316

Bhide, D. (2024). Future of healthcare and AI: Practical insights and diverse perspectives on AI in healthcare project management. *PM World Journal, 13,* 1–11. https://pmworldlibrary.net/wp-content/uploads/2024/05/pmwj138-May2024-Bhide-future-of-healthcare-and-ai-practical-insights.pdf

Communicating the value of medical affairs. (2020, May). A MAPS White Paper. https://medicalaffairs.org/communicating-the-value-of-medical-affairs/

de Bono, E. (1999). *Six thinking hats.* Back Bay Books.

Dylla, H. (2024). The essential roles of medical affairs in healthcare. *Medical Affairs Blog, Catalyst Life Sciences.* https://www.catalystlifesciences.com/blog/the-essential-roles-of-medical-affairs-in-healthcare/

Digital agility and customer-centricity – Imperatives for medical affairs in a post-COVID world. (n.d.). *European Pharmaceutical Review.* https://www.europeanpharmaceuticalreview.com/article/digital-agility-and-customer-centricity-imperatives-for-medical-affairs-in-a-post-covid-world/

Evers, M., Fleming, E., & Ghatak, A., et al. (2014). Pharma medical affairs: 2020 and beyond. McKinsey & Company Insights. https://www.mckinsey.com/industries/life-sciences/our-insights/pharma-medical-affairs-2020-and-beyond

Hippisley-Cox, J., Coupland, C. A., Bafadhel, M., et al. (2024). Development and validation of a new algorithm for improved cardiovascular risk prediction. *Nature Medicine, 30,* 1140–1147. https://doi.org/10.1038/s41591-020-1113-6

How to evaluate technology solutions that enable value-based care. (n.d.). Elace Health. https://www.elacehealth.com/how-to-evaluate-technology-solutions

Gerke, S., Babic, B., Evgeniou, T., & Cohen, G. I. (2020). The need for a system view to regulate artificial intelligence/machine learning-based software as medical devices. *npj Digital Medicine, 3,* 53, 1–4. https://doi.org/10.1038/s41746-020-0262-2

John, S. (2023). How medical affairs can drive patient-centric outcomes. White Paper. ZS Associates. https://www.zs.com/insights/how-medical-affairs-can-drive-patient-centric-outcomes

John, S., & Swant, R. (2024). A patient-centricity framework for medical affairs: Four pillars for improving patient outcomes. White Paper. ZS Associates. https://www.zs.com/insights/a-patient-centricity-framework-for-medical-affairs

Kohles, J., McCall, T., & Williams, L. (2023). Medical affairs strategic planning: Providing a roadmap for success. eCademy Webinar series, MAPS. https://medicalaffairs.org/medical-affairs-strategic-planning/

Lakata, R., Toron, E., Williams, L., et al. (2024). Medical operations: Making the strategy work. eCademy Webinar series, MAPS. https://medicalaffairs.org/medical-operations-making-the-strategy-work/

Lance, L. L. (2024). What is the Scrum methodology? Complete guide to all things Scrum. *The Digital Project Manager*. https://www.thedigitalprojectmanager.com/what-is-scrum-methodology/

Lomas, N. (2019). DeepMind touts predictive healthcare AI breakthrough trained on heavily skewed data. *TechCrunch*. https://techcrunch.com/2019/07/26/deepmind-touts-predictive-healthcare-ai-breakthrough-trained-on-heavily-skewed-data/

MAPS. (2023). *MAPS Midwest Summit Report 2023*. https://medicalaffairs.org/maps-midwest-summit-report-2023/

Medical Affairs Professional Society (MAPS). (2020). *Patient centricity: The future of medical affairs*. https://medicalaffairs.org/patient-centricity-the-future-of-medical-affairs/

Moloto, M., Harmse, A., & Zuva, T. (2020). Impact of agile methodology use on project success in organizations – A SLR. *Software Engineering Perspectives in Intelligent Systems; Conference Paper,* 267–280. http://dx.doi.org/10.1007/978-3-030-63322-6_21

Pracyk, J., Sigmund, W., et al. (2023). The mission, value, and roles of medical affairs in medtech. MAPS. https://medicalaffairs.org/the-mission-value-and-roles-of-medical-affairs-in-medtech/

Project management: How medical affairs receives value every day. (n.d.). CMK Select. https://www.cmkselect.com/project-management-how-medical-affairs-receives-value-every-day/

Rajagopalan, S. (2016, July). TONES: A reference framework for identifying skills and competencies and grooming talent to transform middle management through the field of project management. *International Journal of Markets and Business Systems, 2*(1), 3–24. https://www.inderscience.com/info/inarticle.php?artid=77164

Rajagopalan, S. (2023). 10 ways AI/ML technologies can enhance software quality in the life sciences industry. *PM 360*. https://www.pm360online.com/10-ways-ai-ml-technologies-can-enhance-software-quality-in-the-life-sciences-industry/

Ramadas, S., et al. (2020). Medical affairs 2030: A next-generation operating model. White Paper. Syneos Health. https://www.syneoshealth.com/insights/medical-affairs-2030-a-next-generation-operating-model

Ramadas, S., et al. (2022). The next evolution of medical affairs: The C-Suite. White Paper. Syneos Health. https://www.syneoshealth.com/insights/the-next-evolution-of-medical-affairs-the-c-suite

Rickwood, S., Pereira, F., & Gomez, C. (2022). Attracting investment in clinical development. White Paper. IQVIA. https://www.iqvia.com/library/white-papers/attracting-investment-in-clinical-development

Sheng, J. Q., Hu, P., Liu, X., et al. (2021). Predictive analytics for care and management of

 https://doi.org/10.52202/075831-0002

patients with acute diseases: Deep learning-based method to predict crucial complication phenotypes. *Journal of Medical Internet Research, 23,* e18372. https://doi.org/10.2196/18372

Shepard, K., Kremer, C., & Sundem, G. (2024). The roles, value and practice of medical affairs in the biopharmaceutical and medical technology industries. MAPS 2024. https://medicalaffairs.org/the-roles-value-and-practice-of-medical-affairs-in-the-biopharmaceutical-and-medical-technology-industries/

Terra, J. (2023). Project management in healthcare: Introduction and phases. *UMass Amherst*. https://www.umass.edu/pbs/project-management-in-healthcare

Van Schooenderwoert, N., & Shoemaker, B. (2021). How medical companies are innovating through agile practices. *InfoQ*. https://www.infoq.com/articles/how-medical-companies-are-innovating-through-agile-practices/

Whisman, T. (2023). Large pharma vs small biotech: Which is better, and why? *Medical Affairs Weekly*. https://medicalaffairs.org/large-pharma-vs-small-biotech-which-is-better-and-why/

Author Biography

- **Dr. Premanand Sundivakkam, Ph.D., MBA., BCMAS** is a seasoned board-certified medical affairs professional, recognized for his scientific acumen and medical voice within the pharmaceutical and medical devices industries. His specializations in oncology, hematology, and rare diseases demonstrate his commitment to advancing medical knowledge across diverse fields.

  Dr. Sundivakkam's academic credentials, including a PhD from The Tamil Nadu Dr. MGR Medical University and an MBA from the University of Illinois at Urbana Champaign, have equipped him with a unique blend of scientific expertise and business acumen.

  As a prolific author, Dr. Sundivakkam has contributed to 23 highly cited peer-reviewed journals and served on the editorial board for several internationally acclaimed journals, such as the American Journal of Physiology, PLoS One, and Molecular Pharmacology. His extensive publications underscore his dedication to sharing knowledge and advancing the field of medical affairs.

  In his leadership role at the Medical Affairs Professional Society (MAPS), Dr. Sundivakkam oversees the Midwest Territory and serves in the Working Group on Medical Insights. His leadership extends beyond organizational boundaries, fostering cross-industry partnerships and driving innovation in the field. This is further evidenced by his successful leadership of the 2023 and 2024 Medical Affairs Technology and Innovation Summits.

  Dr. Sundivakkam's professional contributions have been recognized with several awards, including the 2023 Professional Contribution Award by MAPS and the "Top 100 Under 50" award by DiversityMBA. During his academic days, he was an American Heart Fellow, a testament to his longstanding commitment to medical research.

- **Dr. Bratati Ganguly, Ph.D., MBA., BCMASc** is a distinguished leader in Oncology and Medical Affairs, serving as CEO and Founder of Medical Affairs Solutions since 2024. With over a decade of experience, she has driven transformative growth in healthcare. Previously, she excelled as Senior Manager in Medical Affairs at Edwards Life Sciences and Program Manager and Consultant at Syneos Health, overseeing KOL engagement and global projects in Hematology, Solid Tumors, and GI Cancer.

  Dr. Ganguly's career began as a Research Scientist at Brighter Ideas Inc. and progressed through roles such as Pediatric Oncology Scientist at Rutgers Cancer Institute of New Jersey (CINJ) and Marketing Strategy Associate at Rutgers Innovative Ventures. She holds a PhD in Biochemistry from North Dakota State University and an MBA in Marketing and Pharmaceutical Management from Rutgers Business School.

  A prolific author, Dr. Ganguly has 16 peer-reviewed publications and serves on editorial boards for oncology journals. She hosts "The Medical Affairs Dialogues" podcast and co-leads the 2024 MAPS Summit.

  Her accolades include the Joel E. Scholarship Award, Harmon Pearl Scholarship Award, Nicholas and Nancy J. Balaria MB Fellowship, Outstanding Achiever Award from Rutgers Innovative Ventures, Young Investigator Award with a $100,000 grant, Global Outreach Ambassador Award, and Green and Golden Globe Diversity Award. She was recently recognized with the Marquis Who's Who in America 2023 award.

  Supported by her family and mentors, Dr. Ganguly continues to shape Medical Affairs, committed to advancing the field through mentorship and innovation.

- **Dr. Sriram Rajagopalan, PhD., MBA., PMP, PfMP, PgMP** serves as the Global Head of Agile Strategy, Transformation, and Training Services at Inflectra, Assistant Teaching Professor with the College of Professional Studies at Northeastern University, and President of Agile Training Champions doing management training.

  Dr. Rajagopalan has worked as the Vice President of Operational Excellence and Training, Vice President of Proposition Delivery, and Vice President of Program Management Office at Aptus Health. During his tenure as Vice President, he managed a portfolio of programs and projects. Besides establishing the Level 5 Program Management Office (PMO) in Aptus Health and a Project Office in West Notifications Group, Dr. Rajagopalan delivered numerous projects for clients such as eFunds, Northwest Airlines, CVS Pharmacy, Prime Therapeutics, US Airways, Blue Cross Blue Shield, GSK, Novartis, Boehringer-Ingelheim, AstraZeneca, and Depomed. He received the prestigious global PMI Eric Jenett PMI Excellence award in 2017 and was also one of the finalists for the Kerzner award for process excellence in 2012.

  Dr. Rajagopalan has published peer-reviewed scholarly journals and white papers and background papers, on practitioner columns on topics related to middle management transformation through project management. He has created his reference frameworks, such as TONES for middle management transformation, PARAG for product

https://doi.org/10.52202/075831-0002

management and SLEEP for people centric leadership. Dr. Rajagopalan offers training courses and certification prep courses on numerous topics such as Agile, DASSM, PMP, Scrum, Risk, etc.

Dr. Rajagopalan volunteered at PMI Mass Bay Chapter, Agile Alliance, and NAAAP. With PMI Mass Bay Chapter, he served in the board as the Vice President of Marketing and Communication and as the Director of Speakers' Bureau. He served on the board of the BU Agile Innovation Lab. Dr. Rajagopalan founded the "Leadership Matters" initiative with the Northeastern University Leadership program and has been supporting it. Dr. Rajagopalan also actively supports the Global Leadership and Project Management Symposium at Northeastern University and as an advisory board member in the Global Skill Development Council.

Dr. Rajagopalan is passionate about his Projecting Leaders of Tomorrow (PLOT) initiative through which he brings project management and agile principles to youth leadership for children in high schools and to college students in all majors. Dr. Rajagopalan published his book, "Organized Common Sense," to support his PLOT initiative. He has been actively blogging on https://agilesriram.blogspot.com, creates travel vlogs, and is active speaking at numerous conferences such as PMI MassBay, PMI Chicago Agile Conference, InflectraCon, Agile DevOps, Global Conference of Education, Kanban India, and others.

Dr. Rajagopalan earned his bachelor's degree in Electronics and Communication Engineering from the University of Madras in India, a master's (MS) degree in Computer Engineering from Wayne State University, Michigan, a business (MBA) degree in Management from Concordia University, Wisconsin, and a doctorate (Ph.D.) degree in Organization and Management from Capella University, Minnesota. He holds many professional certifications, such as the PfMP, PgMP, PMP, PMI-ACP, PMI-SP, PMI-RMP, SPC, DASSM, PARP, CSP, CSPO, A-CSPO, CSD, CSM, A-CSM, ICC-ACC, SSBB, LSSBB, IT Project+, CSOXP, SCM, SCPO, SCD, SAMC, SCT, and SSMC.

- **Shashi Singh, PMP., CSP., ACP., SAFe SPC4, EMBA** is an Agile Evangelist with over 20 years of experience helping Fortune 500 companies achieve value through leadership and effective project management, processes, and techniques. Adept at providing IT Portfolio Program and Project Management consulting, his company coaches and facilitates customers to achieve sales growth, operational efficiency, and profitability through creating an innovative, Agile culture. He has charted numerous success stories where he has guided clients to achieve quantifiable improvement in predictability, structure, time-to-market, and cultural appetite for continuous improvement.
  Mr. Singh maintains his keen knowledge of trends and techniques through Meetups, Communities of Practice, a wide range of conferences, and by empathically listening to his students, mentees, and clients.

https://doi.org/10.52202/075831-0002

Mr. Singh has earned his Executive MBA from University of IL, Urbana-Champaign and has a B.Tech engineering degree from the renowned Indian Institute of Technology in India.

https://doi.org/10.52202/075831-0002

AUTHOR INDEX